MW00682913

SoccerTowns
Book Two

Andres Varela

Illustrations and Graphic Design by Carlos F. Gonzalez
Co-Producer Germán Hernández
www.soccertowns.com
2014

n Book One, Roundy made some friends at the hospital. They met Teo and started their adventure to learn about the towns where soccer is played in North America.

After their last stop in Houston, they headed North to Kansas City. It was a 12 hour journey.

The group of friends arrived in Kansas City on a beautiful, bright morning. Teo took them to the Liberty Memorial National Historic Landmark. The landmark is a Memorial to the soldiers who died in World War I. Roundy, Shorty, Jersey, Gabe, Ben, Tom, John (Socks), Jerry and Joe (Soccer Cleats) enjoy time in the sun.

5

Kansas City is popularly known as KC.
Over 2 Million people live in the metropolitan area of KC.
A new, soccer-only stadium was built in the year 2011.
The City is famous for its very tasty BBQ!. "We'll go and
eat there later!" says Thomas, "it is my favorite food",
he adds.
KC is located really close to the geographical center
of the United States; that is why some people call it the
heart of America.

Once they learned about KC, the team heads out to
the stadium to play their first game together.
It is an exhibition game, otherwise known as a
"friendly game".

In the changing rooms of the stadium they receive their first uniforms!
They put them on and walk out to the field.
The game they are playing is a "friendly" one and once their game ends, an official game will begin; therefore the stadium is starting to fill up with thousands of soccer fans.

The team comes out to the pitch and can feel the energy of the fans. The TV cameras capture Roundy looking at the big screen.

The team stretches and warm up, following Teo's instructions. His instructions are basic: play a strong defense, do not allow any goals and make sure to get the ball to Jerry and Joe so they can kick it very hard and score from the distance.

It was not a good game, they lost 2-0. The team feels really sad.

In the changing room, Teo shows on the board the mistakes the team made and how not to making them again.

He also motivates the team by saying "In soccer, defeats are part of the game, we need to learn from them and move on. There will be many more games and time to play better, so lift your spirits and let's go eat".

They head out to "The Plaza" which is an area in midtown KC known for nice shops and restaurants. They find a nice BBQ place and enjoy a very good meal. After the meal they all feel better and plan to leave the next day for Chicago, their next stop!

MINNESOTA

WISCONSIN

IOWA

Chicago

TO EAST 80

TO EAST 88

Des Moines

THE PEOPLE OF IOWA WELCOME YOU
Iowa
Fields of Opportunities

WELCOME TO ILLINOIS
THE LAND OF LINCOLN

TO NORTH 35

ILLINOIS

Kansas City

KANSAS

MISSOURI

18

Leaving KC, the team heads up highway I-35 North towards Des Moines, in the State of Iowa.
When they cross the State line between Kansas and Iowa they see the sign "Field of Opportunities".

They get to Des Moines and see the highway signs that say East 80 Davenport Chicago. They head East, following the signs.

Emma says "Wow, there are a lot of big highways everywhere". Teo responds "Yes there are a lot of them, but there are good signs we can follow so we do not get lost. Plus we have one of the best drivers in the world!"

They get to Chicago. It is one of the biggest cities in The United States. In the metropolitan area there are over 9 Million residents, making it the third largest in The United States.

The second tallest building in the country is the Willis Tower, formerly known as the Sears Tower (on the left). Another landmark in the city is the Buckingham Fountain. Another fun place to visit is the Navy Pier, right next to Lake Michigan.

"Let's go there" says Roundy!

The team heads out to the Navy Pier. They get out of the van and Thomas parks it. Tom says "wow, this is a really large lake!". Teo says "Yes, Lake Michigan is the third largest lake in the country".
"Let's go have some fun" says Shorty, while walking to the Pier entrance. The team walk around the Pier, where there are a lot of things to see and do.

They decide to go on the Ferris Wheel, because the view from up there is fantastic!
Teo is the last one to get on and he makes sure everybody is securely wearing their seat belt.
The team is really excited to go up on the wheel and see the skyline of the city and the lake from up above.

Come back and read the next story...

CPSIA information can be obtained
at www.ICGtesting.com
Printed in the USA
LVIC05n1818211214
419836LV00015B/124